Crossing in Oil Palm:

A Manual

Techniques in Plantation Science Series

Series editors:

Brian P. Forster, Lead Scientist, Verdant Bioscience, Indonesia
Peter D.S. Caligari, Science Strategy Executive Director, Verdant Bioscience, Indonesia

About the series:

A series of manuals covering techniques in plantation science that form the essential underlying needs to carry out plantation science.

The series reflects the expertise in Verdant Bioscience that underlies the plantation science activities carried out at the Verdant Plantation Science Centre at Timbang Deli, Deli Serdang, North Sumatra, Indonesia.

Titles available:

1. **Crossing in Oil Palm: A Manual** – Umi Setiawati, Baihaqi Sitepu, Fazrin Nur, Brian P. Forster and Sylvester Dery
2. **Seed Production in Oil Palm: A Manual** – Eddy S. Kelanaputra, Stephen P.C. Nelson, Umi Setiawati, Baihaqi Sitepu, Fazrin Nur, Brian P. Forster and Abdul R. Purbak
3. **Nursery Screening for *Ganoderma* Response in Oil Palm Seedlings: A Manual** – Miranti Rahmaningsih, Ike Virdiana, Syamsul Bahri, Yassier Anwar, Brian P. Forster and Frédéric Breton
4. **Mutation Breeding in Oil Palm: A Manual** – Fazrin Nur, Brian P. Forster, Samuel A. Osei, Samuel Amiteye, Jennifer Ciomas, Soeranto Hoeman and Ljupcho Jankuloski

Crossing in Oil Palm:
A Manual

Umi Setiawati
Verdant Bioscience, Indonesia

Baihaqi Sitepu
Verdant Bioscience, Indonesia

Fazrin Nur
Verdant Bioscience, Indonesia

Brian P. Forster
Verdant Bioscience, Indonesia

Sylvester Dery
Oil Palm Research Institute, Ghana

CABI is a trading name of CAB International

CABI	CABI
Nosworthy Way	745 Atlantic Avenue
Wallingford	8th Floor
Oxfordshire OX10 8DE	Boston, MA 02111
UK	USA
Tel: +44 (0)1491 832111	Tel: +1 (617)682-9015
Fax: +44 (0)1491 833508	E-mail: cabi-nao@cabi.org
E-mail: info@cabi.org	
Website: www.cabi.org	

A catalogue record for this book is available from the British Library, London, UK.

Library of Congress Cataloging-in-Publication Data

Names: Setiawati, Umi, author.
Title: Crossing in oil palm : a manual / Umi Setiawati, Verdant Bioscience, Indonesia [and four others].
Description: Wallingford, Oxfordshire, UK ; Boston, MA : CABI, [2018] | Series: Techniques in plantation science series | Includes bibliographical references and index.
Identifiers: LCCN 2018017371 (print) | LCCN 2018024405 (ebook) | ISBN 9781786395924 (ePDF) | ISBN 9781786395931 (ePub) | ISBN 9781786395917 (pbk : alk. paper)
Subjects: LCSH: Oil palm. | Oil palm--Mutation breeding.
Classification: LCC SB299.P3 (ebook) | LCC SB299.P3 S48 2018 (print) | DDC 633.8/51--dc23
LC record available at https://lccn.loc.gov/2018017371

ISBN-13: 978 1 78639 591 7 (pbk)
          978 1 78639 592 4 (e-book)
          9781786395931 (e-pub)

Commissioning editor: Rachael Russell
Editorial assistant: Emma McCann
Production editor: James Bishop

Typeset by SPi, Pondicherry, India
Printed and bound in the UK by Severn, Gloucester.

Series Foreword – Techniques in Plantation Science

Verdant Bioscience, Singapore (VBS), is a new company established in October 2013 with a vision to develop high-yielding, high-quality planting material in oil palm and rubber through the application of sound practices based on scientific innovation in plant breeding. The approach is to fuse traditional breeding strategies with the latest methods in biotechnology. These techniques are integrated with expertise and the application of sustainable aspects of agronomy and crop protection, alongside information and imaging technology which not only find relevance in direct aspects of plantation practice but also in selection within the breeding programme. When high-yielding planting material is allied with efficient plantation practices, it leads to what may be termed 'intensive sustainable' production. At the same time, the quality of new products is refined to give more specialized uses alongside more commodity-based oil production, thus meeting the market demands of the modern world community, but with a minimal harmful footprint. An essential ingredient in all this is having sound and practical protocols and techniques to allow the realization of the strategies that are envisaged.

To achieve its aims, VBS acquired an Indonesian company called PT Timbang Deli Indonesia, with an estate of over 970 ha of land at Timbang Deli, Deli Serdang, North Sumatra, Indonesia, and the group works under the name of 'Verdant'. A central part of this estate, which will be used for important plant nurseries and field trials, is the development of the Verdant Plantation Science Centre (VPSC), to which the operational staff moved in October 2016. A seed production and marketing facility is now established at VPSC for commercial seed sales and the processing of seed from breeding programmes. The centre comprises specialized laboratories in cell biology, genomics, tissue culture, pollen, soil DNA, plant and soil nutrition, bunch and oil, agronomy and crop protection. Field facilities include extensive nurseries, seed gardens and trials (trial sites are also located at various locations across Indonesia). It is the aim of the company to use its existing and rapidly developing intellectual property (IP) to develop superior cultivars that not only have outstanding yield but also are resistant to both biotic and abiotic stresses, while at the same time meeting new market demands. Verdant not only develops and supplies superior planting materials but also supports its customers and growers with a package of services and advice in fertilizer recommendations and crop protection. This is all part of a central mission to promote green, eco-friendly agriculture.

Brian P. Forster and Peter D.S. Caligari
Lead Scientist and Science Strategy Executive Director
Verdant Bioscience

Contents

Contents

Acknowledgements

The authors are grateful to all the breeding, biotechnology, seed production and marketing teams of Verdant Bioscience for sharing their knowledge and providing helpful advice in preparing this manual. Special mention is given to Heru Rusfiandi and Jennifer Ciomas for sharing their knowledge and providing figures on the reproductive biology of oil palm.

Preface

As noted in the Foreword to this series, a central objective in Verdant's mission is to develop better cultivars of oil palm, rubber and other plantation crops through plant breeding. The higher-yielding the planting material is, the less land that is needed to achieve a specific level of production. Essential to this objective is a wide germplasm base that has been established in the form of live palms, elite lines and pollen collections, which are being positively exploited along with intellectual property (IP) in reproductive biology to develop new cultivars. In addition to this initial stage in the breeding programme, crossing is also used extensively in the production of commercial seed. Crossing activities are therefore central to Verdant's breeding strategy, and these have been developed to maximize success, safeguard purity and provide high-quality products. Protocols developed from these crossing activities provide the basis for this manual. Our target audiences are students and researchers in agriculture, plant breeders, growers and end-users interested in the practicalities of oil palm crossing for breeding and commercial seed production.

Brian P. Forster and Peter D.S. Caligari
Series Editors
February 2018

Introduction

1

Abstract

A brief history of the evolution of the oil palm crop is provided, from domestication and small-scale production in Africa (its centre of origin/diversity) to crop expansion, especially in South-east Asia, in becoming the world's top oil crop. Knowledge of the biology and genetics of oil palm are fundamental in understanding how the crop can be developed through crossing and breeding, and key biological factors such as flowering, outcrossing and genetics are described. A prerequisite for breeding is the crossing of parental lines that carry desirable traits; thus, the germplasm (genetic variation) available to breeders is critical for success. Traditional traits of interest are yield and quality. More recently, disease and pest resistance have become important, as too have novel traits that may be exploited in developing a crop that can be harvested mechanically. Crossing is not only performed to develop and improve elite breeding lines but also is an essential component of commercial seed production.

1.1 History of Oil Palm Cultivation and Crop Facts

Oil palm has the Latin name, *Elaeis guineensis* Jacq.; the genus name is derived from the Greek 'elaion', meaning oil, and the species name indicates its West African origin. The crop was discovered by travellers to Africa in the 15th century, but the first plantings in Indonesia, which led to its rise as the world's pre-eminent oil crop, did not occur until the late 19th century. Large-scale plantations were established in the early 20th century in both Africa and South-east Asia as interest in the crop developed. These initial plantations were composed of Dura palms, which were characterized as having thick-shelled fruits (Fig. 1.1). In the 1920s, the first crosses were made in deliberate attempts to improve the crop through plant breeding, and in the 1950s to 1960s, the more productive Tenera types (a result of crossing Dura with Pisifera) took over as the favoured commercial material

© Umi Setiawati, Baihaqi Sitepu, Fazrin Nur, Brian P. Forster and Sylvester Dery
2018. *Crossing in Oil Palm: A Manual*

Fig. 1.1. Fruit types of oil palm. Fruit cut in half to show: (a) Dura, thick shelled; (b) Tenera, thin shelled with a fibre ring around the shell; (c) Pisifera, no shell, with traces of a fibre ring around the kernel.

in both Africa and South-east Asia. Tenera genotypes are thin shelled, have thick, oil-bearing fruit flesh, and yield 30% more oil than Duras. Thus, crossing became an essential and major component in commercial oil palm seed production as well as in breeding.

A review of the oil palm crop is given by Corley and Tinker (2008). Oil palm is grown in the humid tropics, between 20° latitudes north and south of the equator, and covers over 8.5 million hectares (Mha) worldwide. The crop is highly profitable and is grown both on large-scale plantations and by smallholders (Sayer *et al.*, 2012). Ripe fruit bunches are harvested continually and sent to local mills for oil extraction. Oil palm fruits provide both crude palm oil (CPO) and palm kernel oil (PKO), extracted from the fruit flesh (mesocarp) and kernel (endosperm), respectively. CPO is made up of palmitic (43%), oleic (39%), stearic (5%) and other fatty acids (Siew, 2002), and is a major source of provitamin A and vitamin E (Barcelos *et al.*, 2015). PKO is a high-quality oil containing lauric (up to 50%), myristic (15%) and other essential fatty acids (Sambanthamurthi *et al.*, 2000). Since oil palm is harvested continually, CPO represents a relatively stable commodity compared to annual oil crops. The main CPO-producing countries are Indonesia (53% of global production) and Malaysia (38%); the largest consumers are India (28% of the market), Europe (22%) and China (22%).

1.2   History of Oil Palm Breeding

Although the oil palm has been used by humans in West Africa since ancient times, its commercial exploitation is of recent origin. With the realization of the economic potential of the crop, concerted efforts towards its improvement started as early as the turn of the 20th century in both West and Central Africa. An account of the history of the development of the various breeding programmes in Africa has been explained briefly by Hartley (1967, 1988). This early effort includes the works of the Institut National pour l'Etude Agronomique du Congo belge (INEAC) and two major West African programmes: namely, the West African Institute for Oil Palm Research (WAIFOR) in Nigeria, Ghana and Sierra Leone, and the Institut de Recherche pour Les Huiles et Oleagineux (IRHO) in Cote d'Ivoire and Benin.

The work in Nigeria was started by E. Smith, of the Nigerian Department of Agriculture, by exploiting the genetic variation in the oil palm population of Calabar, Aba, Nkwele (Umuahia) and later Ufuma in natural groves of the ten eastern Nigeria regions. He planted 800 oil palms at Calabar in south-eastern Nigeria. The yield and bunch composition of these plants (arising from seeds of open-pollinated bunches collected from local wild grove palms) was collected from 1922 to 1928. From this he selected nine Duras (fruit with thick-shelled kernels) and ten Teneras (thin shelled). Twelve of the selected palms were self-pollinated to form the Calabar F1 (first) generation, and the progenies were planted in four stations of the Department of Agriculture between 1930 and 1935 (Ogba, Umudike, Ibadan and Nkwelle). These stations supplied extension work seeds (EWS) and seeds for experimentation and performance testing (Barbosa and Chinchilla, 2003).

In Indonesia, similar work was initiated by the Algemene Vereniging van Rubberplanters ter Oostkust van Sumatra (AVROS) and based on four palms received from Africa and planted as ornamental palms in Bogor Botanical Gardens, Java, Indonesia. Seed from these four palms was widely distributed throughout Indonesia as ornamentals (often used to line roads in rubber plantations), but was later used to supply oil palm estates (from 1911 onwards). Selection programmes started at various centres in Indonesia from 1920 onwards and gave rise to a breeding population of Duras generally referred to as 'Deli Dura' (Hardon and Turner, 1967; Hardon, 1976).

The superior oil content of the Tenera (T) over the Dura (D) fruit types led to the distribution of T×T seed for commercial planting in Congo in the early 1930s. This was not a successful strategy as, in 1938, it was found that as much as 25% of progeny were sterile. The underlying genetics of this was explained by Beirnaert and Vanerweyen (1941), who described clearly the simple one gene (*Sh*) inheritance of the shell thickness character (examples given in Fig. 1.2), which dismissed the prevailing theory in Congo at the time that Tenera types were a 'degenerating form' of oil palm. They explained that female sterility (of Pisifera types in the progeny) could be prevented by D×T or T×D crosses and that these could replace T×T production in which 25% of the progeny were expected to be female sterile Pisifera types. Working in Nigeria, Africa, they demonstrated the predicted 1:2:1 segregation ratios (D:T:P) in Calabar Tenera selfings (described in Hartley, 1967). The understanding that shell thickness was controlled by a single gene exhibiting simple Mendelian inheritance revolutionized the oil palm industry, as it was understood that 100% Tenera progeny could be produced from D×P crosses (although Pisiferas are female sterile, there is no problem with pollen production and they may be used as male parents). Commercial Tenera production from D×P crosses including Deli Dura mother palms as parents was initiated and became established in the 1950s (Corley and Tinker, 2003).

Parents	Dura ♀	Dura ♂	Tenera ♀	Tenera ♂	Dura ♀	Pisifera ♂
Phenotype	Thick shell	Thick shell	Thin shell	Thin shell	Thick shell	No shell
Genotype	Sh/Sh	Sh/Sh	Sh/sh	Sh/sh	Sh/Sh	sh/sh
Cross	X ↓		X ↓		X ↓	
Progeny	100% Dura (Sh/Sh)		25% Dura (Sh/Sh) 50% Tenera (Sh/sh) 25% Pisifera (sh/sh)		100% Tenera (Sh/sh)	

Fig. 1.2. Expected segregation ratios in progeny from various fruit type crosses (D×D, T×T and D×P).

1.3   Biology and Genetics of Oil Palm

Oil palm (*E. guineensis*) is a long-lived perennial. It has a single apical meristem and does not branch. After germination, there is a 3-year juvenile stage before inflorescences appear. These are produced in each leaf axil and are either male or female (monoecious); the first are normally male. Oil palm is an outbreeding species, and pollination is effected predominantly by the weevil, *Elaeidobius kamerunicus*, which depends on oil palm inflorescences to complete its life cycle. Oil palm inflorescences are large and typically give rise to bunches containing 100–4,000 fruits, depending on palm age. Fruits mature at about 150 days after pollination, turning from black to red, and begin to fall out of the bunch when ripe. The fruit is a drupe (stone fruit) composed of a fleshy mesocarp and a central kernel protected by a shell (endocarp). The kernel contains the products of fertilization: embryo and endosperm. In the wild, the mesocarp of fallen fruit rots or is eaten, leaving behind the kernel; germination takes place in favourable conditions by the emergence of a seedling shoot and root through the germ pore (operculum) in the shell. Generally, one seedling is produced per seed, but up to three may occur.

Oil palm is diploid, with 16 pairs of chromosomes, but is thought to have evolved from an ancient tetraploid species as there is extensive genome duplication (Singh *et al.*, 2013). It is highly heterozygous, owing to its outbreeding reproductive system. Oil palm can be self-pollinated artificially but suffers from inbreeding depression. Genetic maps of the oil palm genome have been developed (Billotte *et al.*, 2005) and the genome has been sequenced (Singh *et al.*, 2013); thus, genetic markers may be deployed in screening for genes of interest in progeny from deliberate crossings.

1.4 Pollen Biology of Oil Palm

Although pollen is vital in plant reproduction, and specifically in breeding and seed production, there are few reports on the pollen biology of oil palm. This is particularly concerning for oil palm as pollen is not only essential for breeding (crossing) but also for commercial success in the production of Tenera seed from Dura × Pisifera crosses.

Pollen development takes place in the anthers of male florets in oil palm. Pollen grains are the end products of meiosis in which chromosomes recombine, reassort and are packaged into gametic cells that develop into pollen grains. Male meiosis occurs in specialized cells in the anther, known as pollen mother cells (microsporocytes), these are diploid ($2n$ = 32 chromosomes), and results in uninucleate cells (microspores), which are haploid (n = 16 chromosomes). The single nucleus of each microspore then divides (first mitosis) to produce binucleate microspores. These two nuclei have separate destinies; one is vegetative, the other generative. At maturity, oil palm pollen is binucleate (in most plant species, the mature pollen is trinucleate as the generative cell divides to produce two sperm during pollen development). The binucleate nature of oil palm pollen is thought to be a factor that allows oil palm pollen to be stored (more than 10 years). In oil palm, the division of the generative nucleus does not occur until after the pollen has reached a receptive stigma on a female floret (pollination) and after pollen germination; it takes place in the pollen tube as it grows down the style towards the embryo sac of the female floret (see Fig. 1.5).

A detailed account of pollen development, including the timing of events, is given by Nasution *et al.* (2009); a simplified scheme is given in Fig. 1.3. Pollen grain development occurs inside the anther and can be divided into the following stages based on cytological examination:

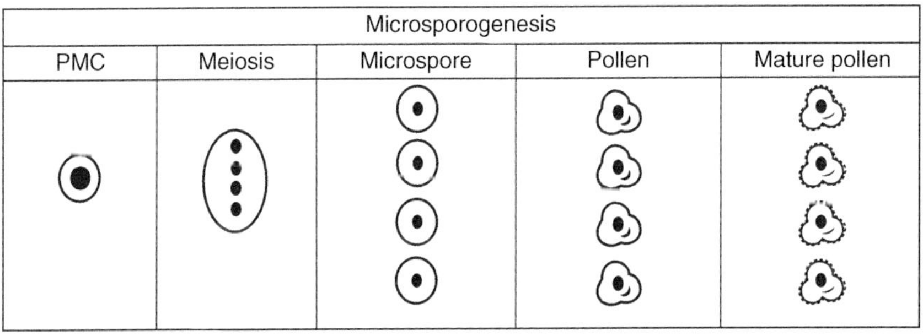

Microsporogenesis				
PMC	Meiosis	Microspore	Pollen	Mature pollen

Fig. 1.3. A simplified schematic of pollen development (microsporogenesis) in oil palm, from the pollen mother cell (PMC) to mature pollen.

1. Pollen mother cell (PMC). This is a specialized cell that undergoes meiosis. The pollen mother cell, like the parent palm, is diploid with 32 (16 pairs of) chromosomes.

2. Tetrad. These four cells (microspores) are the product of meiosis of one pollen mother cell. Each of these microspores contains one haploid nucleus consisting of 16 single chromosomes. The four cells are clumped together to form a tetrad.
3. Empty microspore. The microspores have separated from the tetrad into individual entities and appear to be empty (highly vacuolated with little cytoplasm).
4. Starch-filled microspore. The cytoplasm of the microspores develops and fills up with starch, giving a granular appearance when viewed under a light microscope.
5. Late uninucleate. The nucleus becomes more apparent as it prepares to divide (first mitosis).
6. Binucleate. The microspore contains two asymmetric haploid nuclei: one vegetative nucleus and one generative nucleus. The vegetative nucleus appears larger but less dense than the generative nucleus.
7. Mature pollen. At maturity, the oil palm pollen is binucleate, is triangular in shape and has a protein-rich exterior and a germination pore. In nature, it is viable for about 6 days after anthesis.

Pollen production is affected by palm development, health of the palm and the environment. Male inflorescences are usually the first to appear once the juvenile palm becomes mature (3–5 years after germination). Stress conditions such as drought promote the development of male inflorescences. Factors that affect pollen development, anthesis and viability include: humidity; temperature; and rainfall. Oil palm pollen may be collected, dried and stored, as described in Chapter 3 of this manual (see also Ekaratne and Senathirajah, 1982).

1.5   Embryo Sac Biology of Oil Palm

The embryo sacs of flowering plants have been classified into various types (Davis, 1967). A key feature in this classification is the number of megaspores that give rise to gametic cells of the embryo sac. Four megaspores cells are produced by meiosis from each megasporocyte (megaspore mother cell). However, the fate of these cells varies: some may die, whereas others may produce gametic cells. For example, in the tetrasporic class, all four megaspores contribute to the gametic cells of the embryo sac; in the monosporic class, three of the four megaspores abort, leaving one to divide mitotically to produce all the gametic cells of the embryo sac.

The development of the embryo sac of oil palm is not fully understood. Light microscopic studies of oil palm embryo sacs described them variously as polygonum type (monosporic; Kajale and Ranade, 1953) and adoxa type (tetrasporic; de Poerck, 1950). Both studies agree that the embryo sac of oil palm is composed of eight gametic (haploid) cells but disagree on the origins. The polygonum type is generally accepted as more likely, although the real situation may be mixed. In both cases, embryo sac development begins with

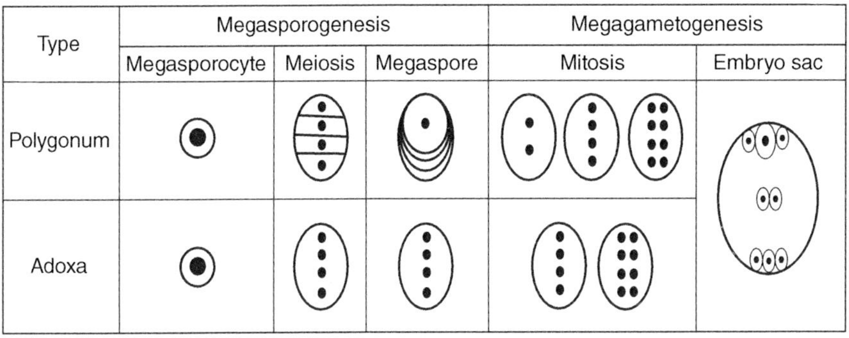

Type	Megasporogenesis			Megagametogenesis	
	Megasporocyte	Meiosis	Megaspore	Mitosis	Embryo sac
Polygonum					
Adoxa					

Fig. 1.4. Simplified schemes of polygonum (monosporic) and adoxa (tetrasporic) embryo sac development leading to an eight-celled embryo sac.

the megasporocyte (embryo sac mother cell), which is a specialized cell that undergoes meiosis to produce four megaspores. In the polygonum type, three of these perish, leaving just one (monosporic); this undergoes two rounds of mitosis to produce an eight-celled embryo sac. In the adoxa type, all four megaspores survive and undergo one round of mitosis to produce a similar eight-celled embryo sac. In both types, the eight cells are organized into: one egg cell; two synergid cells; a central cell composed of two fused cells; and three antipodal cells. A simplified schematic of polygonum and adoxa embryo sac formation is given in Fig. 1.4. The embryo sac is positioned inside the ovary of the carpel (female floret), which also comprises the style and the stigma.

1.6 Fertilization in Oil Palm

Pollination is the arrival of pollen on the stigma. Pollen may be transmitted by various modes: wind; water; insects; animals, etc. Oil palm is naturally insect pollinated, predominantly by a weevil (see Section 1.3 above), but may be artificially pollinated, as described in Chapter 6 of this manual. Once the pollen grain arrives on a receptive stigma of a female floret, it begins to germinate (see Chapter 4 of this manual for artificial pollen germination *in vitro*). A pollen tube emerges from the germination pore of the pollen grain and grows into the style towards the embryo sac. The vegetative nucleus appears to act as a chaperone in taking the generative nucleus down the pollen tube (Nasution *et al.*, 2009), similar to other species, Brassicas for example (Dumas *et al.*, 1985). While in the pollen tube, the generative cell divides (second mitosis) to produce two sperm, one of which is associated with the vegetative nucleus; this union is known as the male germ unit (Tian *et al.*, 1998). The pollen tube grows into the micropyle, which leads to the embryo sac. The male germ unit is thought to have a role in positioning the two sperm for double fertilization, with one fertilizing the egg cell and the other fertilizing the central cell (Russell *et al.*, 1990; Yu and Russell, 1994).

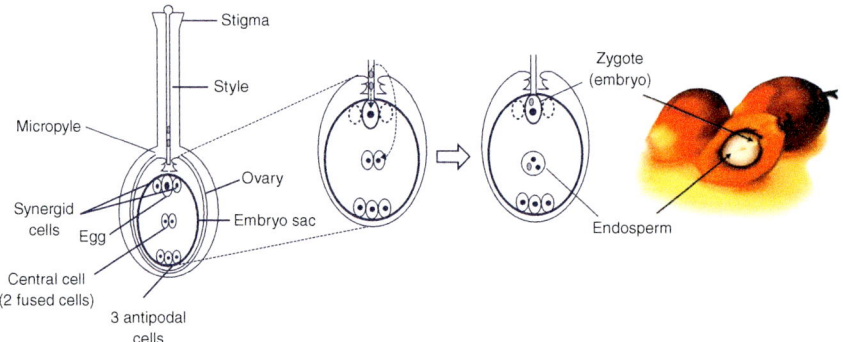

Fig. 1.5. Events after pollen arrives on the stigma (pollination) leading to double fertilization (embryo and endosperm formation) in the kernel of oil palm fruits.

After double fertilization, the egg cell becomes a zygote (diploid) and develops into an embryo, while the central cell becomes the endosperm (triploid). The endosperm becomes a food reserve for seed development and seed germination.

The events from pollination to fertilization of oil palm are described in detail in a timed cytological study by Nasution *et al.* (2009); the main features are summarized above in Fig. 1.5.

1.7   Germplasm

Commercial seed production is based on Dura × Pisifera crosses, which produce the desired thin-shelled Tenera. The predominant parental lines have been Deli Duras and AVROS Pisiferas. Crop improvement through breeding has been limited by the genetic variation contained in these elite parental gene pools. In order to make progress in breeding, it became necessary to provide breeders with more genetic variation, and thus germplasm collections from wild, landrace and cultivated materials in West Africa (the centre of diversity) have been carried out. WAIFOR, Nigeria, was one of the first to do this. In recent years, major oil palm breeding companies have joined collecting expeditions in West Africa (Okyere-Boateng *et al.*, 2008; Sapey *et al.*, 2012).

Oil palm germplasm is conserved as living palm trees or as pollen. The former requires large land areas but may be maintained for up to 100 years, as oil palm is long-lived; pollen may be stored for up to 20 years vacuum packed in –20°C freezers. Another approach suggested for oil palm is to cryopreserve germplasm at –198°C in liquid nitrogen (Grout *et al.*, 1983).

1.8   Target Traits

As with most crops, yield is the most important trait for oil palm breeding. Yield per land area is of particular importance as the oil palm industry is under pressure to be environmentally friendly and sustainable (limit expansion and to conserve rainforest biodiversity).

Other traits of interest include:

- Short stature: reduced height enables longer plantation life.
- Precocity: early flowering brings early economic returns.
- Resistance to disease: wilt in Africa and *Ganoderma* in South-east Asia.
- Mechanized harvesting: fruit colour, long bunch stalk, fruit abscission.
- Oil quality: develop specialized oils for specific end-users.

Target traits for oil palm improvement and methods in plant breeding have been reviewed by Forster *et al.* (2018).

1.9   Commercial Crossing

Tenera oil palm has become the predominant commercial fruit type. Tenera seed is produced by commercial seed companies by crossing Dura (thick shelled) with Pisifera (no shell) genotypes. Pisifera palms suffer from female sterility and are therefore used as male (pollen) parents in commercial production. Pisifera pollen is therefore collected and stored ready for use in crossing. The shell thickness trait is controlled by a single gene: *Sh*, with Dura being homozygous dominant (*Sh/Sh*), Pisifera homozygous recessive (*sh/sh*) and Tenera heterozygous (*Sh/sh*). Seed production is a specialized and lucrative business: in 2014, Indonesian oil palm seed producers sold 102,826,918 seeds at an average cost of about US$0.8/seed. The oil palm industry is therefore dependent on quality controlled crossing procedures.

References

Barbosa, R. and Chinchilla, C. (2003) ASD oil palm germplasm from Nigeria. *ASD Oil Palm Papers* 26, 33–44.

Barcelos, E., de Almeida Rios, S., Cunha, R.N.V., Lopes, R., Motoike, S.Y. *et al.* (2015) Oil palm natural diversity and the potential for yield improvement. *Frontiers in Plant Science* 6, 190. Available at: http://doi.org/10.3389/fpls.2015.00190 (accessed 5 March 2018).

Beirnaert, A. and Vanerweyen, R. (1941) Contribution à l'étude génétique et biométrique des variétés d'*Elaeis guineensis* Jacquin. *INEAC Série Scientifiqué* 27, 1–101.

Billotte, N., Marseillac, N., Risterucci, A.M., Adon, B., Brottier, P. *et al.* (2005) Microsatellite-based high density linkage map in oil palm (*Elaeis guineensis* Jacq.). *Theoretical and Applied Genetics* 110, 754–765.

Corley, R.H.V. and Tinker, P.B. (2003) *The Oil Palm*, first edn. John Wiley & Sons, London.

Corley, R.H.V. and Tinker, P.B. (2008) *The Oil Palm*, fourth edn. John Wiley & Sons, London.

Davis, G.L. (1967) *Systeatic Embryology of the Angiosperms*. John Wiley & Sons, London, 528 pp.

de Poerck, R.A. (1950) Contribution a l'etude du Palmier a huile African *Elaeis guineensis* Jacq. *Oleagineaux* 5, 754–765.

Dumas, C., Knox, R.B. and Gaude, T. (1985) The spatial association of the sperm cells and vegetative nucleus in the pollen grain of *Brassica*. *Protoplasma* 124, 168–174.

Ekaratne, S.N.R. and Senathirajah, S. (1982) Viability and storage of pollen of oil palm (*Elaeis guineensis* Jacq.). *Annals of Botany* 51, 661–668.

Forster, B.P., Setiawati, U., Sitepu, B., Kelanaputra, E.S., Nur, F. *et al.* (2018) Oil palm breeding. In: Campos, H. and Caligari, P.D.S. (eds) *Genetic Improvement of Tropical Species*, Chapter 6. Springer, Switzerland.

Grout, B.W.W., Shelton, K. and Pritchard, H.W. (1983) Orthodox behaviour of oil palm seed and cryopreservation of the excised embryo for genetic conservation. *Annals of Botany* 52, 381–384.

Hardon, J.J. (1976) Oil palm breeding – Introduction in oil palm research. *Developments in Crop Science* 1, 98–107 ref. 2pp. Record No. 19771653683.

Hardon, J.J. and Turner, P.D. (1967) Observations on natural pollination in commercial plantings of oil palm (*Elaeis guineensis*) in Malaya. *Experimental Agriculture* 3, 105–116.

Hartley, C.W.S. (1967) *The Oil Palm*. Longmans, Green and Co Ltd, London, 706 pp.

Hartley, C.W.S. (1988) *The Oil Palm*, third edn. Longman, London.

Kajale, L.B. and Ranade, S.G. (1953) The embryo sac of *Elaeis guineensis* Jacq. A reinvestigation. *Journal of the Indian Botanical Society* 32, 101–107.

Nasution, O., Rusfiandi, H., Sitorus, A.C., Forster, B.P., Nelson, P.C. and Caligari, P.D.S. (2009) Cytological studies of pollen development in oil palm (*Elaeis guineensis* Jacq.). *Proceedings of PIPOC 2009*, Kuala Lumpur, 9–12 November 2009, AP47, pp. 954–961.

Okyere-Boateng, G., Dwarko, D.A., Kaledzi, P.D. *et al.* (2008) Collection, conservation and evaluation of the disappearing oil palm (*Elaeis guineensis* J) landraces in Ghana. *International Journal of Pure and Applied Sciences and Technology* 1, 18–31.

Russell, S.D., Cresti, M. and Dumas, C. (1990) Recent progress on sperm characterization in flowering plants. *Physiologia Plantarum* 80, 669–676.

Sambanthamurthi, R., Kalyana, S. and Tan, Y. (2000) Chemistry and biochemistry of palm oil. *Progress in Lipid Research* 39, 507–558.

Sapey, E., Adusei-Fosu, K., Agyei-Dwarko, D. and Okyere-Boateng, G. (2012) Collection of oil palm (*Elaeis guineensis* Jacq.) germplasm in the northern region of Ghana. *Asian Journal of Agricultural Sciences* 4, 325–328.

Sayer, J., Ghazoul, J., Nelson, P. and Boedhihartono, A.K. (2012) Oil palm expansion transforms tropical landscapes and livelihoods. *Global Food Security* 1, 114–119.

Siew, W.L. (2002) Palm oil. In: Gunstone, F.D. (ed.) *Vegetable Oil in Food Technology: Composition, Properties and Uses*. Wiley-Blackwell, New Jersey, pp. 25–58.

Singh, R., Ong-Abdullah, M., Low, E.T.L., Manaf, M.A.A., Rosli, R. *et al.* (2013) Oil palm genome sequence reveals divergence of interfertile species in old and new worlds. *Nature* 500, 335–339.

Tian, H.Q., Zhang, Z. and Russell, S.D. (1998) Isolation of the male germ unit: organization and function in tobacco (*Nictoniana tabacum* L.). *Plant Cell Reports* 18, 142–147.

Yu, H.-S. and Russell, S.D. (1994) Male reproductive cell development in *Nicotiana tabacum*: male germ unit associations and quantitative cytology during sperm maturation. *Sexual Plant Reproduction* 7(6), 324–332.

Health and Safety Considerations   **2**

Abstract

All field and laboratory operations should have standard health and safety protocols. These may vary according to local requirements and standards. Some equipment will also come with instructions on proper use, which may involve training, including health and safety issues. Failure to abide by these can result in accidents and personal injury (serious and minor); neglect of health and safety issues may incur penalties such as fines or cessation in field and laboratory activities. Guidelines in health and safety issues relating to crossing in oil palm are given below.

2.1   Health and Safety in the Field

Oil palm crossing is carried out in the field and a major health and safety issue is the height of palm trees. Inflorescences are located in the canopy of the palm trees and these can reach a height of 15 m after 20 years. It is recommended that young, short palms are used whenever possible, but if older, taller trees are used for either pollen collection and/or crossing, then several safety issues must be taken into consideration.

Equipment needed for tall palm trees

- Ladders are used for climbing tall oil palm trees. Ladders are usually made of iron or other hard material.
- Harnesses, helmets, boots, etc., are used for protection against accidents; for example, falling when climbing ladders.
- Specialist clothing to protect against general abrasion with plant parts.
- Training.

Other general considerations

- Sharp knives are used to open the spathe (outer cover of the inflorescence).
- Dangerous endemic insects and animals; for example, mosquitos, snakes.
- Standard operating procedures (SOPs).
- Working alone.
- Emergency procedures, first aid box.
- Wear a mask; hazardous chemicals (formalin and insecticides) are used (sprayed) in the isolation of female inflorescence to safeguard against uncontrolled pollinations. Face protection is therefore required to prevent inhalation and eye and skin contact.
- Wear gloves; hazardous chemicals (formalin and insecticides) are used (sprayed) in the isolation of female inflorescence to safeguard against uncontrolled pollinations. Hand protection is therefore required to prevent skin contact. Any contact with the hazardous chemicals should be followed immediately by washing and first aid.

2.2   Health and Safety in the Laboratory

Crossing of oil palm requires various laboratory procedures, especially in the treatment, storage and preparation of pollen. Good laboratory practices are therefore required.

- Put on a laboratory coat before entering the laboratory and remove it when leaving the laboratory. This provides protection to yourself and the samples you are working with in the laboratory, and protects people outside the laboratory against contamination from laboratory materials.
- Wear appropriate clothing; skin exposure should be minimized. Field clothing should be removed before entering the laboratory.
- Training in making ampoules – glassware, pulling glass (necking) and flaming fire hazard.
- Be aware of emergency procedures: firefighting, emergency exits, emergency telephone numbers, and the location of fire extinguishers and first aid/first aiders.
- Be aware of hazards relating to chemicals used in the laboratory and read their Material Safety Data Sheet (MSDA information is available on the Internet), which provides information on health and safety, first aid, fire and explosion risks, disposal, how to clean up spillage, handling and storage.
- Be aware of SOPs that have been developed for your laboratory (for example, use heat-resistant gloves when placing equipment in or removing it from an oven) or which should be developed (for example, for waste disposal).

More information on safe procedures in the laboratory are given by Barker (2005).

Reference

Barker, K. (2005) *At the Bench: A Laboratory Navigator*. Cold Spring Harbor Press, New York.

Pollen Collection and Storage   3

Abstract

Shed pollen of oil palm is binucleate, it has two cells (one vegetative and one generative), and can live for about a week *in situ*. However, fresh pollen can be collected from flowering male inflorescences, purified and stored for over 20 years in cold, desiccated and vacuumed conditions. The processes involved are described. Male inflorescences are isolated in bags prior to flowering, to ensure pollen purity, and treated to prevent insects bringing in pollen from other sources. Specialized isolation bags are used that contain a smaller pollen-collecting bag and a window to observe development. Male inflorescences are harvested during the late stages of anthesis and placed in a hot room to encourage the remaining florets to open and shed pollen. The collected pollen is purified by sieving, drying and vacuum packing in ampoules and stored ready for use.

Pollen is collected from isolated male inflorescences. These may involve tall trees that require ladders and safety equipment. Hazardous chemicals and sharp knives are used in field operations; see Chapter 2 of this manual for guidance.

3.1 Tools

- Gloves – used for handling stock chemical solution. Generally, gloves are not needed when using working solutions, as chemicals are diluted.
- Chisel – used for removing the palm spines and bending the fronds downwards to gain access to inflorescences and bunches.
- Knife – used to open the spathe (outer protection of the inflorescence).
- Safety equipment for tall palms: ladders, boots, harness, helmet.
- The pollen box is sterilized by spraying with 96% ethanol and heating to more than 100°C for 15 min. It is used to mix pollen with talcum powder in sterile conditions.

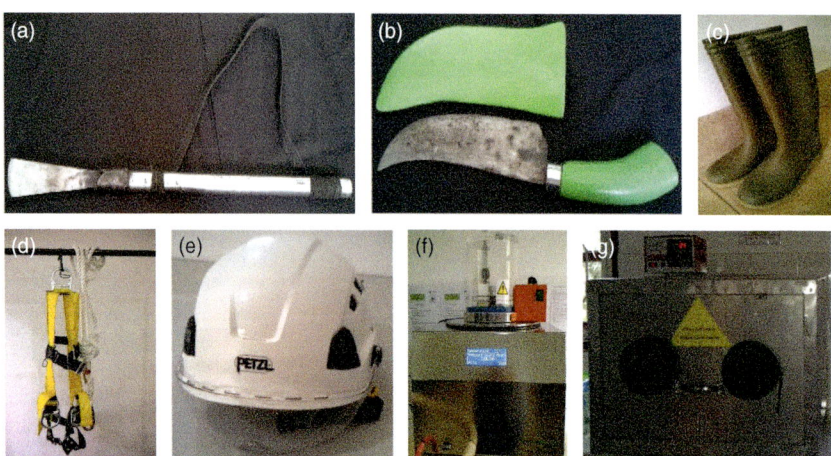

Fig. 3.1. Pollen collection tools: (a) chisel; (b) knife; (c) boots; (d) safety harness; (e) helmet; (f) freeze-dryer; (g) pollen box.

- Freeze-dryer – used to dehydrate the pollen and to inhibit the growth of unwanted organisms through water removal, thus allowing for convenient long-term storage.
- Vacuum pump – used to evacuate the air during pollen ampouling.
- Fridge – used to store the collected pollen that has been ampouled (Fig. 3.1).

3.2 Materials

- Modified isolation bags (fitted with a pollen-collecting bag) – used in the isolation of the male inflorescence.
- Used isolation bags – previously used bags are utilized as secondary protection in male flower isolation.
- Formalin solution (2 ml/l) – used to kill any pollen and insect present.
- Ethanol 96% – utilized to sterilize tools needed for inflorescence isolation.
- Cotton wool – used to place granular insecticide utilized in preventing insect ingress via the stalk.
- Silica gel – used in maintaining a low pollen moisture content.
- Insecticide granules – used for killing unwanted insects.
- Rubber rope – used to bind the isolated male inflorescence to prevent fracture due to wind direction.
- Sieve – 100 mm mesh – used to separate pure pollen from other materials, anthers for example.
- Ampoules – 0.5 and 2.5 ml – used for pollen storage (Fig. 3.2).

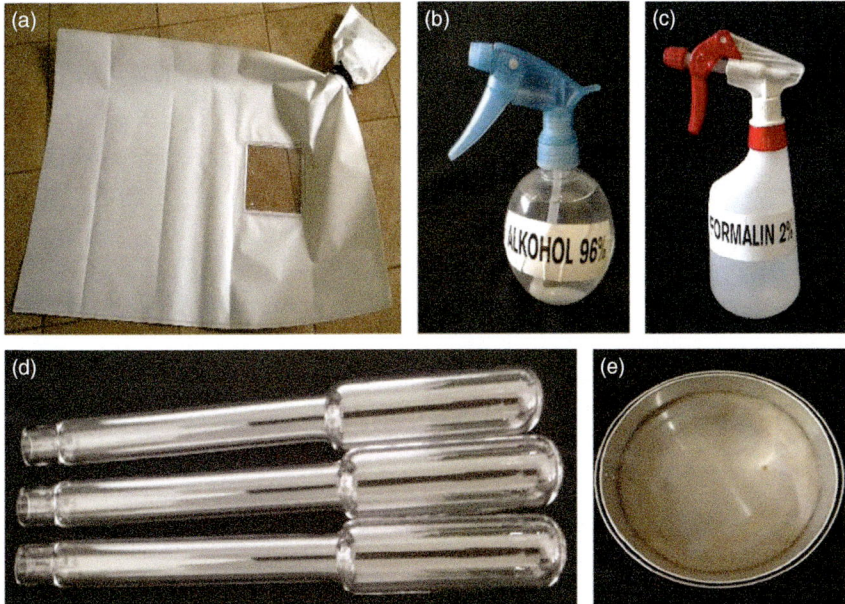

Fig. 3.2. (a) Modified isolation bag; (b) 96% ethanol in hand sprayer; (c) 2% formalin in hand sprayer; (d) ampoule glass; (e) pollen sieve.

3.3   Methods

Step 1

Inflorescence isolation usually takes place when one-third of the spathe has opened. Frond spines should be removed (cleaned) and fronds bent downwards to provide easier working conditions. The spathes are cut open with a knife and sprayed with formalin solution to kill any insects present (Fig. 3.3).

Step 2

Double terylene bags are used to cover and to isolate the male inflorescence. Ideally, the inner bag should be new and modified with the insertion of a pipe leading to a small plastic bag in one corner for pollen collection. The second bag is normally a previously used terylene bag. Insecticide granules are placed inside a cotton wool wad and placed round the inflorescence stalk to prevent the introduction of pollen (carried by insects) from other sources. Old/used terylene bags are applied as a protective outer layer and tied with rubber rope. More than 8 days are then required for complete anther dehiscence (pollen shedding). If the male inflorescence undergoes anthesis (pollen shedding) in less than 9 days, the male inflorescence must be rejected as the legitimacy of the pollen cannot be guaranteed. In field conditions, pollen remains viable for at least 6 days after release from the anther (Hardon and Turner, 1967; Ekaratne and Senathirajah, 1983) (Fig. 3.4).

Fig. 3.3. Step 1. (a) Male inflorescence ready to be isolated; (b) cut open the spathes; (c) spray with 2% formalin.

Fig. 3.4. Step 2. (a) Bag modification; (b) cotton wool is placed around the stalk; (c) bag is tied with a rubber rope; (d) isolated male inflorescence; (e) outer cover is a used bag; (f) bag is tied with a rubber rope; (g) write down the palm identity and isolation date.

Step 3

The isolated male inflorescence is normally harvested when one-third of the flowers of the inflorescence are open and shedding pollen (at anthesis). Flowers start to open from the base of the spikelet, and usually all flowers will open within 1–2 days (Fig. 3.5).

Step 4

The harvested male inflorescence (shedding pollen) is stored by hanging upside down in a hot room (34–39°C) for about 18 h to allow all remaining florets to open and shed their pollen into the inner bag, which is fitted with a collecting bag (Fig. 3.6).

Step 5

After storage in the hot room, the male inflorescence is hand-beaten to channel the pollen downwards into the collecting bag (Fig. 3.7).

Step 6

The collected pollen is sieved in a sterile box to separate the pollen from debris. The interior of the pollen box is sprayed with 96% alcohol, to sterilize

Fig. 3.5. Step 3. (a) Anthesis of a male inflorescence inside bag and (b) exposed.

Fig. 3.6. Step 4. Drying of a pollen-shedding male inflorescence.

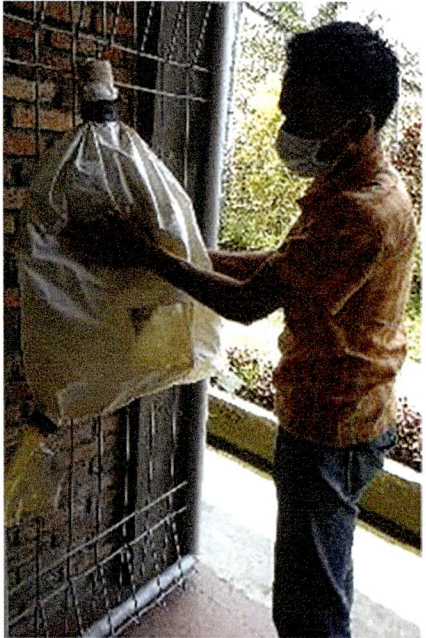

Fig. 3.7. Step 5. Hand-beat the dried male inflorescence.

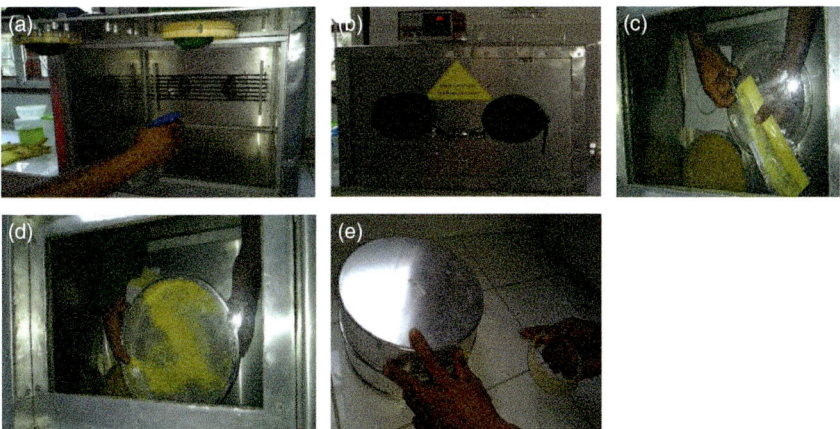

Fig. 3.8. Step 6. (a) Spray with 96% ethanol; (b) pre-heat the box; (c) open the plastic bag and pour the pollen on to a sieve; (d) pollen sieving; (e) dry the purified pollen in the sieve above silica gel.

surfaces, and allowed to dry. The box is then heated to 100°C for around 15 min. The box is then cooled down using an air conditioner and/or fan. When the oven has cooled down, the pollen (in a plastic bag) is placed inside. The pollen is sieved to separate the pollen from debris. The purified pollen is wrapped in thin paper and then dried in the sieve above silica gel for about 18 h (Fig. 3.8).

Step 7

The ampouling process is conducted in a sterilized box/chamber. The purified pollen sample is placed into glass ampoules and the mouth of the ampoule plugged with cotton wool. Two sizes of ampoule are used: 0.5 ml filled with a maximum of 0.6 g of pollen, and 2.5 ml filled with a maximum of 4 g of pollen. A calibrated tube is used to measure the amount of purified pollen. The ampoule is then stacked into an acrylic bell jar and freeze-dried for about 45 min. Clean the top of the ampoule by rubbing with the cotton wool, cut off the top with the cotton wool bung. Heat the ampoule in the centre and pull on both sides carefully. Proper training is required to do this work. Heat-resistant gloves are not needed by trained workers as the worker needs to feel how hot the ampoule becomes to prevent burning the pollen. Stack the ampoules in a drying manifold and vacuum for 45 min. Seal the ampoule by welding while they are still stacked (Fig. 3.9).

Step 8

The vacuumed pollen samples are stored in a freezer below 0°C for more than 5 years. For long-term storage, store at −86°C (Fig. 3.10).

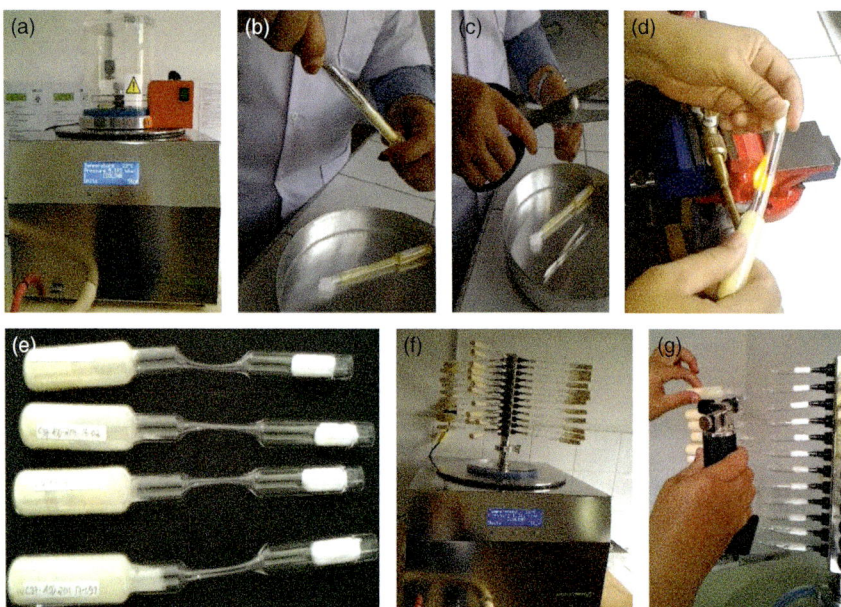

Fig. 3.9. Step 7. (a) Freeze-drying pollen; (b) clean the top of the glass by rubbing with cotton wool; (c) cut off the top with the cotton wool bung; (d) heat the ampoule and pull to form a neck; (e) ampoules resulting from the necking process; (f) vacuuming; (g) sealing.

Fig. 3.10. Step 8. Pollen is stored in a freezer.

References

Ekaratne, S.N. and Senathirajah, S. (1983) Viability and storage of pollen of the oil palm, *Elaeis guineensis* Jacq. *Annals of Botany* 51, 661–668.

Hardon, J.J. and Turner, P.D. (1967) Observations on natural pollination in commercial plantings of oil palm (*Elaeis guineensis*) in Malaya. *Experimental Agriculture* 3, 105–116.

Pollen Viability Testing

4

Abstract

Pollen viability is essential for success in pollination. Fresh pollen shed from flowering male inflorescences has the highest viability, and it is this which is harvested and stored (Ekaratne and Senathirajah, 1983). Viability tests are carried out prior to processing pollen to ensure the harvested pollen has high viability, and again after storage to monitor viability decline with time in storage. Viability testing provides information on pollen quality and when resampling is required. It also determines the ratio of pollen:talc to be used to maximize pollination success with minimal impact on pollen stores. There are many methods that test pollen viability; the standard one described here is based on pollen germination, it is simple and inexpensive. Pollen with high viability is vital in promoting high seed set and acceptable seed numbers for oil palm breeding and commercial seed production, and various techniques have been developed to test for pollen viability (Nasution *et al.*, 2009). Here, we provide a simple method using pollen culture and a light microscope.

4.1 Tools

- Microscope fitted with ×40 objective lens – used to examine pollen germination/viability.
- Microscope slides and coverslips – used to stage the pollen for viewing in a microscope.
- Mounted needle – used to take up pollen sample.
- Incubator (35–37°C) – used to incubate pollen preparation for the germination process.
- Hotplate stirrer – used to dissolve sucrose in hot distilled water (Fig. 4.1).

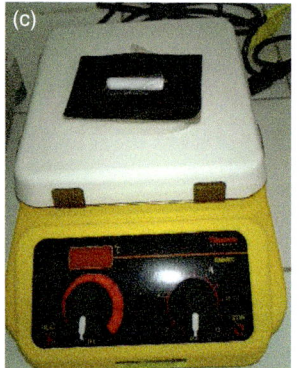

Fig. 4.1. Pollen viability testing tools: (a) microscope (with a ×40 objective lens); (b) incubator; (c) magnetic hotplate stirrer.

4.2   Reagents

- Sucrose or household sugar 10% w/v.
- Boric acid 5% w/v (5 g boric acid dissolved in 95 ml distilled water).
- Distilled water (Fig. 4.2).

4.3   Methods

Step 1

The first step is the preparation of the pollen germination medium: 10 g of sucrose is dissolved in 100 ml of hot distilled water. The solution is cooled and 15 drops of 5% boric acid added (Fig. 4.3).

Step 2

Pollen samples are collected using the tip of a needle and stirred into a drop of the culture medium on a microscope slide. A coverslip is then applied (Fig. 4.4).

Step 3

The sample is incubated at 34–37°C for about 3 h and then examined under a microscope at ×40. The number of germinated (viable) pollen grains is counted along with ungerminated (presumed dead) pollen grains in several fields of view until a minimum of 150 pollen grains have been scored. The percentage of germinated pollen (viability) is then calculated (Fig. 4.5).

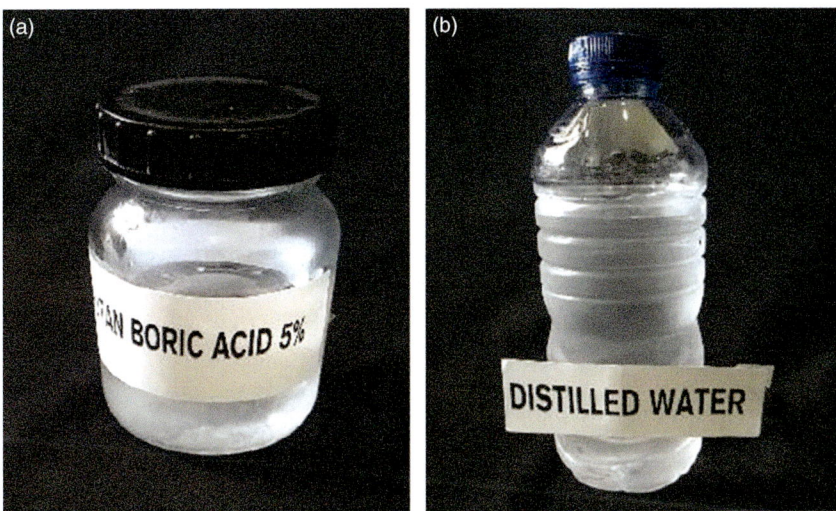

Fig. 4.2. Reagents: (a) 5% boric acid solution; (b) distilled water.

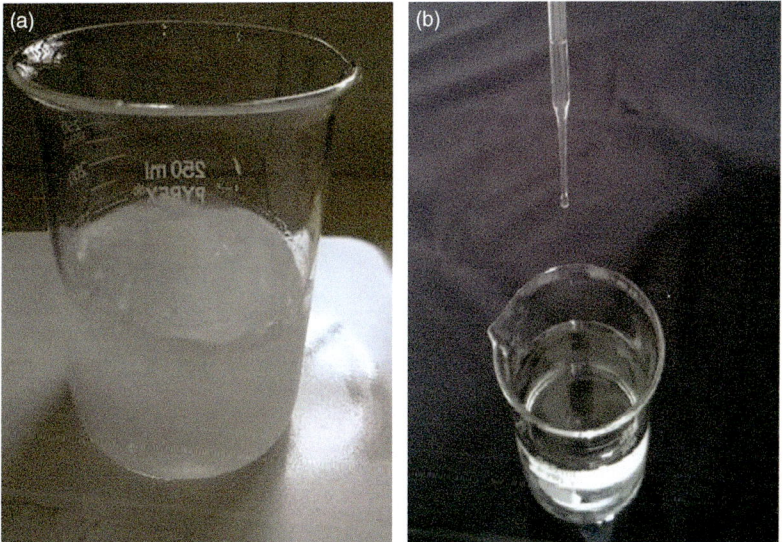

Fig. 4.3. Step 1. (a) Dissolve sucrose in distilled water; (b) drops of 5% boric acid are added to the sucrose solution.

Other viability tests of oil palm pollen using various stains (fluorescein diacetate, DAPI (4′,6-diamidino-2-phenylindole) and aceto-orcein) involving fluorescence microscopy have been developed (Nasution *et al.*, 2009).

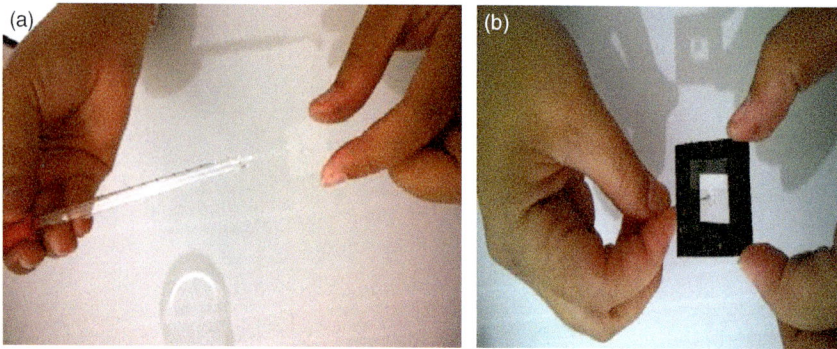

Fig. 4.4. Step 2. (a) Place a drop of liquid medium on to a coverslip; (b) stir the pollen sample into the medium on the coverslip.

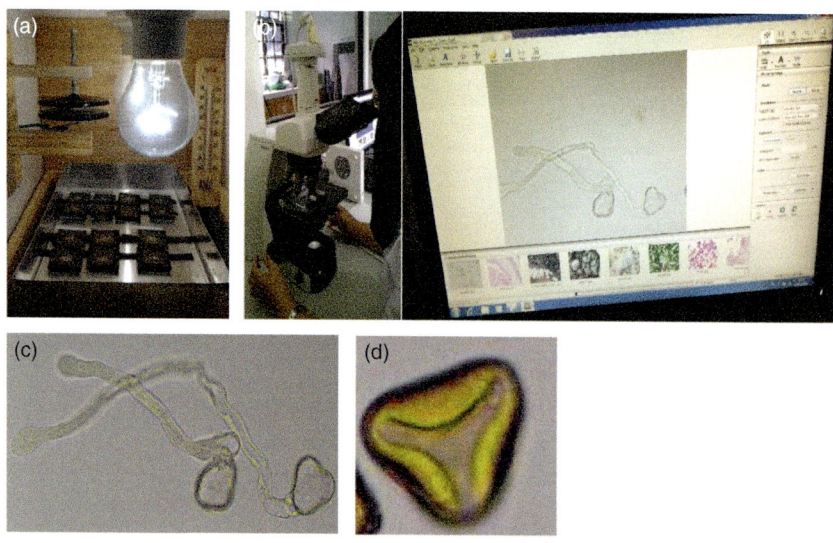

Fig. 4.5. Step 3. (a) Incubation of the samples; (b) examine under a microscope; (c) germinated (live) pollen grains; (d) dead (empty) pollen grain.

References

Ekaratne, S.N. and Senathirajah, S. (1983) Viability and storage of pollen of the oil palm, *Elaeis guineensis* Jacq. *Annals of Botany* 51, 661–668.

Nasution, O., Setiawati, U., Sitorus, A.C., Forster, B.P., Nelson, S.P.C. and Caligari, P.D.S. (2009) Determination of oil palm fertility using staining techniques. *Proceedings of PIPOC 2009*, AP49.

Isolation of the Female Inflorescence

5

Abstract

Care is required in isolating female bunches to prevent damage. The developing inflorescence should be treated with fungicides and insecticides to prevent disease that may result in reduced viability or abortion of the inflorescence. Specialized isolation bags are equipped with a window to inspect and monitor development and through which subsequent operations (spraying and pollination) are performed. The isolation process involves insect proofing, as insects, particularly weevils, carry unwanted pollen and ruin controlled crossing. Instructions are given on the procedures used in isolating the female inflorescence that maximize success in artificial crossing.

The procedures involve the use of hazardous chemicals (insecticides and fungicides) and may involve climbing tall palm trees. Guidance on health and safety procedures is given in Chapter 2 of this manual.

5.1 Tools and Materials

- Terylene bag (new) – used to cover the isolated inflorescence.
- Recycled terylene bag – used to cover the isolated inflorescence as a second protective layer.
- Formalin solution (2 ml/l) – used to kill pollen and any insects that are present.
- 96% ethanol – used to sterilize tools.
- Cotton wool – used as a holder for the granular insecticide needed to prevent the ingress of insects via the stalk.
- Silica gel – used to maintain a low pollen moisture content.
- Insecticide granules – used to kill insects entering the bag via the stalk.
- Face mask – used when handling insecticide.
- Rubber rope – used to bind the isolated male inflorescence to prevent fracture due to wind damage.
- Small scissors – used to make a hole in the isolation bag's window to allow the entry of a popper pollination pipe.

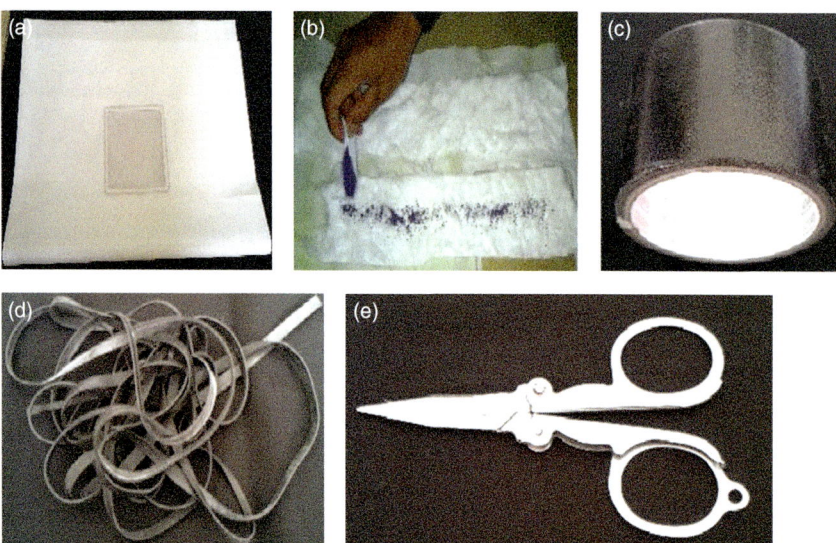

Fig. 5.1. Tools to isolate female inflorescence: (a) terylene bag; (b) cotton wool with insecticide granules (blue); (c) linen tape; (d) rubber rope; (e) small scissors.

- Linen tape – used to close the hole in the isolation bag's window after pollination to prevent other contamination.
- Chisel – used to bend the frond downward to allow easy access to work with the inflorescence.
- Gloves – used when handling insecticide (Fig. 5.1).

5.2   Reagents

- 2% formalin solution – used to kill pollen and insects.
- Insecticide such as 'Baygon', 'Mortin', 'Hits', etc. – used to kill insects (Fig. 5.2).

5.3   Methods

Step 1

Palms are inspected for the emergence of female flower bunches. These are noted, and their development is monitored prior to isolation (Fig. 5.3).

Step 2

The outer and inner spathes are removed from the selected female inflorescence using a knife, to expose the developing florets. The inflorescence is then sprayed with 2% formalin to kill any pollen and insects present. A wad

Fig. 5.2. Reagents to isolate female inflorescence: (a) 2% formalin in a hand sprayer; (b) aerosol insecticide.

Fig. 5.3. Step 1. Female inflorescence.

of cotton wool containing insecticide is then tied to the inflorescence stalk; this kills any insects that may carry unwanted pollen (Fig. 5.4).

Step 3

After the removal of the spathes, the female inflorescence is covered with two terylene bags. The inner bag is a new bag, the outer bag is normally

Fig. 5.4. Step 2. (a) Spray with 2% formalin; (b) tie a cotton wool wad containing insecticide granules to the inflorescence stalk.

Fig. 5.5. Step 3. (a) Cover with a new terylene bag (inner bag); (b) cover with a used terylene bag (outer bag); (c) record the identity of the palm and the date of isolation.

a recycled old bag. The bags are equipped with a window for monitoring development and through which pollination is effected. At least 9 days are needed for the female florets to become receptive, fully opened flowers. If the female inflorescence becomes receptive in less than 9 days, it is rejected, as uncontrolled pollination may have occurred (Fig. 5.5).

Pollination

6

Abstract

Pollination is effected through a window in the specialized isolation bag. This reduces the risk of contamination (pollen, disease and insects). Prior to pollination, the area around the isolation bag is sprayed to kill insects, as these can carry unwanted pollen. Pollen is normally mixed and diluted with talcum powder to provide a volume easy to handle. This mixture is blown on to receptive female inflorescences using specialized equipment that has been surface sterilized. The isolation bag is also shaken to spread the pollen over the entire inflorescence. Labels are then attached showing pollination dates and parental genotypes. The isolation bags may be removed once fruit set is established and harvested when fruits are mature.

Pollination is normally carried out in the morning, when nectar is visible on female florets. Mature pollen grains of oil palm are binucleate, containing one vegetative and one generative nucleus. When the pollen grain arrives on a receptive stigma of a female inflorescence, it begins to germinate. The pollen tube grows down the style, carrying with it the two nuclei; the generative nucleus divides to produce two sperm cells, which are deposited into the ovary, where they effect fertilization (see Chapter 1, Section 1.4 of this manual). For more details on oil palm pollen development, see Nasution *et al.* (2009).

6.1 Tools

- Pollen mixing box – used as a sterile place for mixing pollen.
- Pollen blower (popper) – used to blow pollen into the isolation bag.
- Labels – needed to keep a record and monitor events.
- Small scissors – needed to make a hole in the isolation bag's window to allow the entry of a popper pipe for pollination.
- All tools used for pollination are sterilized using 96% ethanol (Fig. 6.1).
- Linen tape – used to close the hole in the isolation bag's window after pollination to prevent other contamination.

© Umi Setiawati, Baihaqi Sitepu, Fazrin Nur, Brian P. Forster and Sylvester Dery
2018. *Crossing in Oil Palm: A Manual*   41

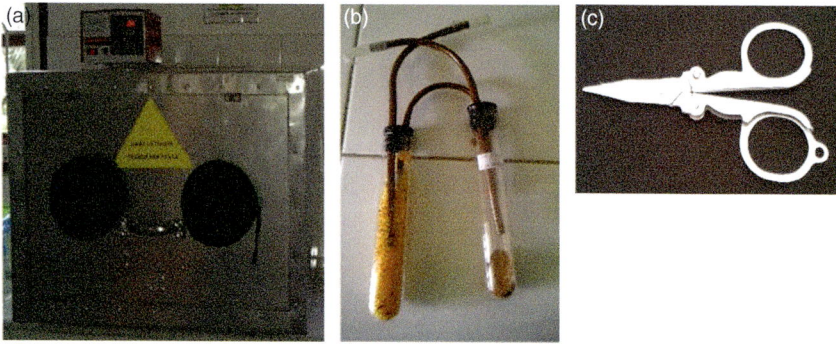

Fig. 6.1. Tools: (a) pollen mixing box; (b) pollen blower equipment; (c) small scissors.

6.2   Reagents

- 96% ethanol – used to sterilize tools.
- Silica gel – used to maintain a low pollen moisture content.
- Insecticide such as 'Baygon', 'Mortein', etc. – used to kill insects.
- Talcum powder – used as a pollen carrier to increase the area covered by pollen (Fig. 6.2).

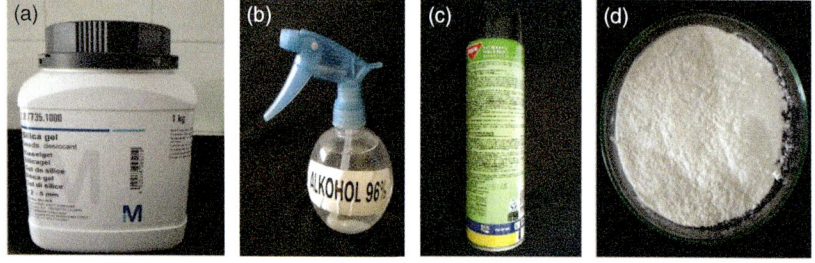

Fig. 6.2. Reagents: (a) silica gel; (b) 96% ethanol in hand sprayer; (c) aerosol insecticide; (d) talcum powder.

6.3   Methods

Step 1

Prior to pollination, the area around the isolated female inflorescence is sprayed with insecticide to kill insects that may carry unwanted pollen (Fig. 6.3).

Step 2

Pollen for pollination is mixed with talcum powder, usually in a ratio of 0.03–0.1 g of pollen (for pollen with >80% viability) with 1 g of talcum

Fig. 6.3. Step 1. (a) Receptive female inflorescence viewed through the bag window; (b) spray the bag with aerosol insecticide; (c) cut the top of the outer bag; (d) spray the inside of the outer bag.

powder (Table 6.1; Fig. 6.4). This is done in a sterilized pollen mixing box to prevent contamination.

Table 6.1. Amount of pollen needed for pollinations based on viability.

Viability (%)	Pollen used per pollination (g)	Number of pollinations from 1 g of pollen
>80	0.100	10
61–80	0.125	8
41–60	0.200	5
21–40	0.250	4
<20	1.000	1

Fig. 6.4. Step 2. Pollen blower equipment.

Step 3

Before pollination is carried out, all tools (including the pollinator's hands) have to be sterilized with 96% ethanol. Pollination is achieved by blowing the pollen + talc mixture into the inner isolation bag by piercing a hole in the bag window. It is normally carried out once, in the morning, to achieve maximum fruit set. The isolation bag is shaken after each delivery of pollen by the blower, to help distribute the pollen over the whole inflorescence. The bag is resealed by covering the bag's window with linen tape and labelling it with the cross details (female and male parents and dates) (Fig. 6.5).

Fig. 6.5. Step 3. (a) Wipe the bag window with 96% ethanol; (b) pierce the bag window with small scissors; (c) blow the pollen and talcum mixture on to the receptive inflorescence; (d) cover the hole with linen tape; (e) label the outer bag with details of the cross and the pollination date.

Step 4

At 21–25 days post-pollination, the isolation bags are removed.

Step 5

The bunch is inspected for fruit set 60–67 days post-pollination.

Step 6

A pollinated bunch is usually harvested 150 days after pollination or when the bunch has produced at least one loose fruit. The ripeness period for each population is different and is also influenced by weather. The ripe bunch is then harvested by cutting the stalk. Using a plastic net to cover the ripe bunch before harvesting is an alternative to protect loose fruit from dropping and becoming mixed with the loose fruit from another bunch on the floor. Place the harvested bunch into a gunny sack and deliver to seed production, where the seed is processed as described in Kelanaputra *et al.*, 2018.

References

Kelanaputra, E.S., Setiawati, U., Sitepu, B., Nur, F., Forster, B.P. and Dery, S. (2018) *Oil Palm Seed Production: A Manual. Techniques in Plantation Science.* Forster, B.P. and Caligari, P.D.S. (eds) CAB International, Wallingford, UK (in press).

Nasution, O., Rusfiandi, H., Sitorus, A.C., Forster, B.P., Nelson, S.P.C. and Caligari, P.D.S. (2009) Cytological studies of pollen development in oil palm (*Elaeis guineensis* Jacq.). In: *Proceedings of Agriculture, Biotechnology and Sustainability Conference*, PIPOC 9–12 November 2009, Kuala Lumpur, Malaysia, Vol 2. Malaysian Palm Oil Board, Kuala Lumpur, pp. 954–961.

Commercial Tenera Production

7

Abstract

Contemporary oil palm cultivars have a thick, oil-bearing mesocarp and thin-shelled fruits, that is, a Tenera phenotype, and this is controlled genetically by the shell thickness gene, *Sh*. The Tenera genotype is heterozygous (*Sh/sh*) and is produced by crossing a thick-shelled female Dura (*Sh/Sh*) with a no-shell male Pisifera (*sh/sh*). Commercial production of Teneras needs to be controlled rigorously. Crossing is carried out as described above, but additional measures are taken to ensure purity of the product; these are described below.

Dura palms are used as mother palms, and the Pisiferas are used as father (pollen) palms. Pollen from selected commercial father palms is collected on a routine basis and stored ready for use. The crossing process is the same as that described above but may include more rigorous procedures to ensure quality of the commercially produced seed; for example, high pollen viability and legitimacy testing.

The seed production programme requires selected Duras and Pisiferas. The Dura selection is based on progeny performance by D×P progeny trials (Forster *et al.*, 2017). The Pisifera selection is based on individual Pisiferas through Pisifera testing trials.

Most oil palm seed producers in the world use Deli Duras as mother palms, since they guarantee offspring with superior bunch and fruit characteristics. There are some research stations in Africa that do not use Deli Dura lines as female parents, because the major objectives are to obtain *Fusarium oxysporum* wilt resistance and drought tolerance.

Blank pollination

Blank pollination is performed during training and as a quality control measure in commercial seed production. Instead of using a pollen:talc mix, the receptive female inflorescence is 'pollinated' with talcum only (no pollen), but the pollinator is unaware which samples contain pollen and which are blank. The method is used to check whether there is any unwanted pollen reaching the receptive isolated female inflorescence. If stray pollen does ingress and achieve fertilization, the resulting seed will be a mixture of types – depending on the genotype of the stray pollen.

The causes of unwanted stray pollen are:

1. The tools used, or the operator's hands, were not sterile.
2. There were some odd male flowers in the female inflorescence.
3. The bag was not well sealed and insects carrying pollen entered the closed bag.

Methods in seed production, the next step after crossing, are described in Forster *et al.* (2017) and in another manual in this series (*Seed Production in Oil Palm: A Manual*; Kelanaputra *et al.*, 2018).

References

Forster, B.P., Setiawati, U., Sitepu, B., Kelanaputra, E.S., Nur, F. *et al.* (2017) Oil palm (*Elaeis guineensis*). In: Campos, H. and Caligari, P.D.S. (eds) *Genetic Improvement of Tropical Crops*. Springer, Chapter 8.

Kelanaputra, E.S., Nelson, S.P.C., Setiawati, U., Sitepu, B., Nur, F. *et al.* (2018) *Seed Production in Oil Palm: A Manual*. *Techniques in Plantation Science*. Forster, B.P. and Caligari, P.D.S. (eds). CAB International, Wallingford, UK (in press).

Index

Page numbers in **bold** type refer to figures and tables.

CABI – who we are and what we do

This book is published by **CABI**, an international not-for-profit organisation that improves people's lives worldwide by providing information and applying scientific expertise to solve problems in agriculture and the environment.

CABI is also a global publisher producing key scientific publications, including world renowned databases, as well as compendia, books, ebooks and full text electronic resources. We publish content in a wide range of subject areas including: agriculture and crop science / animal and veterinary sciences / ecology and conservation / environmental science / horticulture and plant sciences / human health, food science and nutrition / international development / leisure and tourism.

The profits from CABI's publishing activities enable us to work with farming communities around the world, supporting them as they battle with poor soil, invasive species and pests and diseases, to improve their livelihoods and help provide food for an ever growing population.

CABI is an international intergovernmental organisation, and we gratefully acknowledge the core financial support from our member countries (and lead agencies) including:

Ministry of Agriculture People's Republic of China

Australian Government
Australian Centre for International Agricultural Research

Agriculture and Agri-Food Canada

Ministry of Foreign Affairs of the Netherlands

Schweizerische Eidgenossenschaft
Confédération suisse
Confederazione Svizzera
Confederaziun svizra
Swiss Agency for Development and Cooperation SDC

Discover more

To read more about CABI's work, please visit: **www.cabi.org**

Browse our books at: **www.cabi.org/bookshop**, or explore our online products at: **www.cabi.org/publishing-products**

Interested in writing for CABI? Find our author guidelines here: **www.cabi.org/publishing-products/information-for-authors/**